NATURAL DISASTERS
IN INFOGRAPHICS

EnviroGraphics

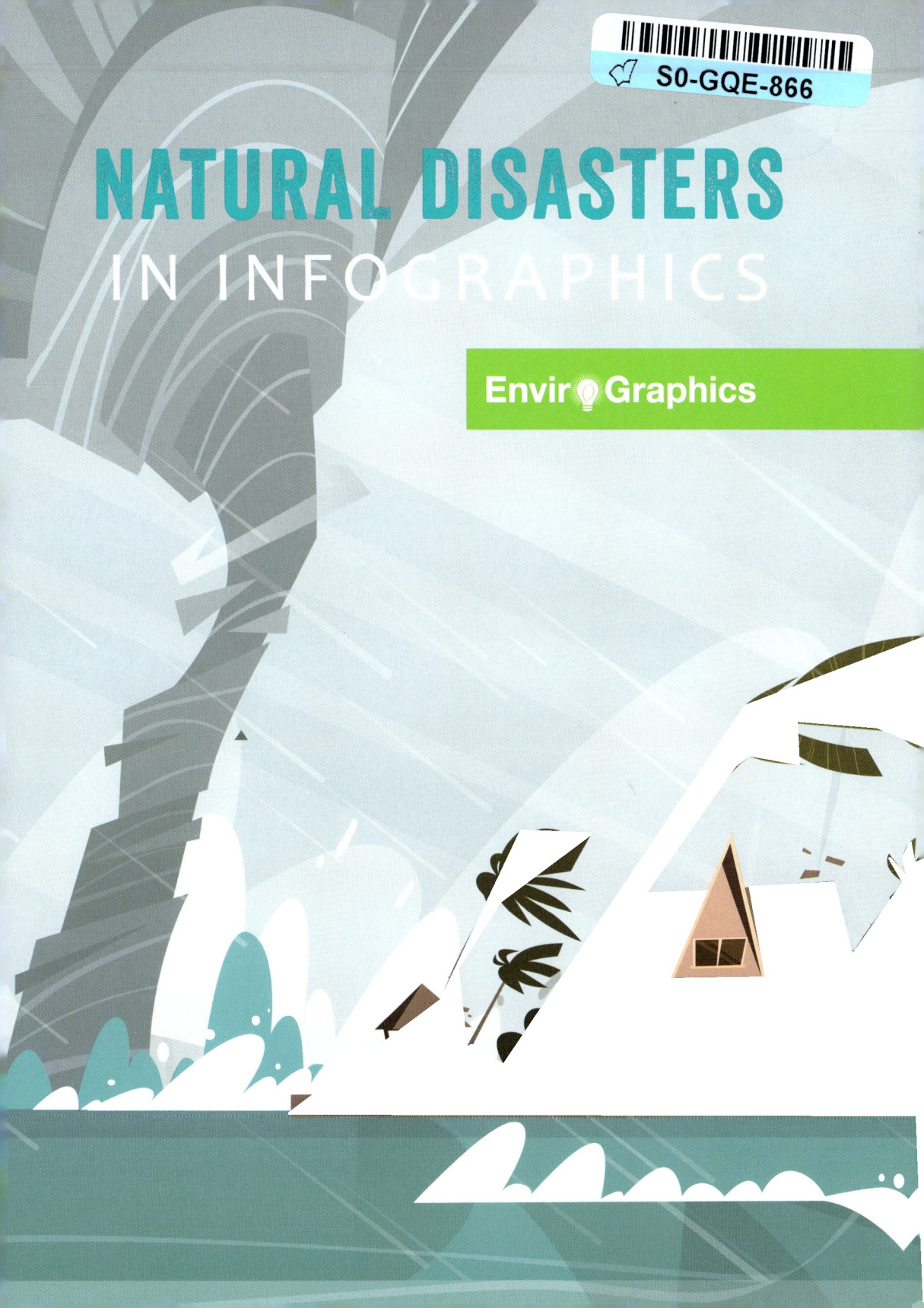

CHERRY LAKE PRESS

Published in the United States of America by Cherry Lake Publishing Group
Ann Arbor, Michigan
www.cherrylakepublishing.com

Reading Adviser: Marla Conn, MS, Ed., Literacy specialist, Read-Ability, Inc.
Photo Credits: ©Clker-Free-Vector-Images/Pixabay, cover; ©OpenClipart-Vectors/Pixabay, cover; ©Shutterstock, cover; ©Shutterstock, 1; ©Shutterstock, 5; ©OpenClipart-Vectors/Pixabay, 6; ©Webflippy/Pixabay, 6; ©Shutterstock, 6; ©Republic of Peru/Wikimedia, 6; ©CryptoSkylark/Pixabay, 7; ©OpenClipart-Vectors/Pixabay, 7; ©Shutterstock, 7; ©Shutterstock, 9; ©Shutterstock, 10; ©Shutterstock, 11; ©Shutterstock, 12; ©OpenClipart-Vectors/Pixabay, 14; ©Shutterstock, 14; ©Shutterstock, 15; ©Shutterstock, 16; ©Shutterstock, 17; ©Shutterstock, 18; ©Shutterstock, 20; ©Shutterstock, 21; ©Shutterstock, 22; ©Clker-Free-Vector-Images/Pixabay, 24; ©OpenClipart-Vectors/Pixabay, 24; ©Webflippy/Pixabay, 24; ©Clker-Free-Vector-Images/Pixabay, 25; ©OpenClipart-Vectors/Pixabay, 25; ©Shutterstock, 25; ©Shutterstock, 26; ©Shutterstock, 28; ©Shutterstock, 30

Cherry Lake Press is an imprint of Cherry Lake Publishing Group.

Library of Congress Cataloging-in-Publication Data has been filed and is available at catalog.loc.gov

Cherry Lake Publishing Group would like to acknowledge the work of the Partnership for 21st Century Learning, a Network of Battelle for Kids. Please visit http://www.battelleforkids.org/networks/p21 for more information.

Printed in the United States of America
Corporate Graphics

TABLE OF CONTENTS

What Are Natural Disasters?

Natural disasters are major weather events that are caused by Earth. The disasters can cause great damage. Natural disasters happen in areas where people live. Homes and businesses can be destroyed. People can be injured or even die. They are much more **severe** than normal storms.

[21ST CENTURY SKILLS LIBRARY]

TORNADOES

FLOODS

EARTHQUAKES

HURRICANES

WILDFIRES

TSUNAMIS

History of Natural Disasters

October 1871

☠ **1,200–2,500**

Peshtigo Fire

United States 🇺🇸

January 2010

☠ **300,000**

Haiti Earthquake

Haiti 🇭🇹

May 1970

☠ **22,000**

Huascarán Avalanche

Peru 🇵🇪

6

February 1972

☠ **4,000**

Iran Blizzard

Iran 🇮🇷

July 1931

☠ **1–4 MILLION**

China Floods

China 🇨🇳

November 1970

☠ **500,000**

Bhola Cyclone

Bangladesh 🇧🇩

2019, Live Science

Research shows that some places are more **vulnerable** to natural disasters than others. This map shows the 10 riskiest places in the United States for these events. This is based on the number of disasters from 1953 to 2013.

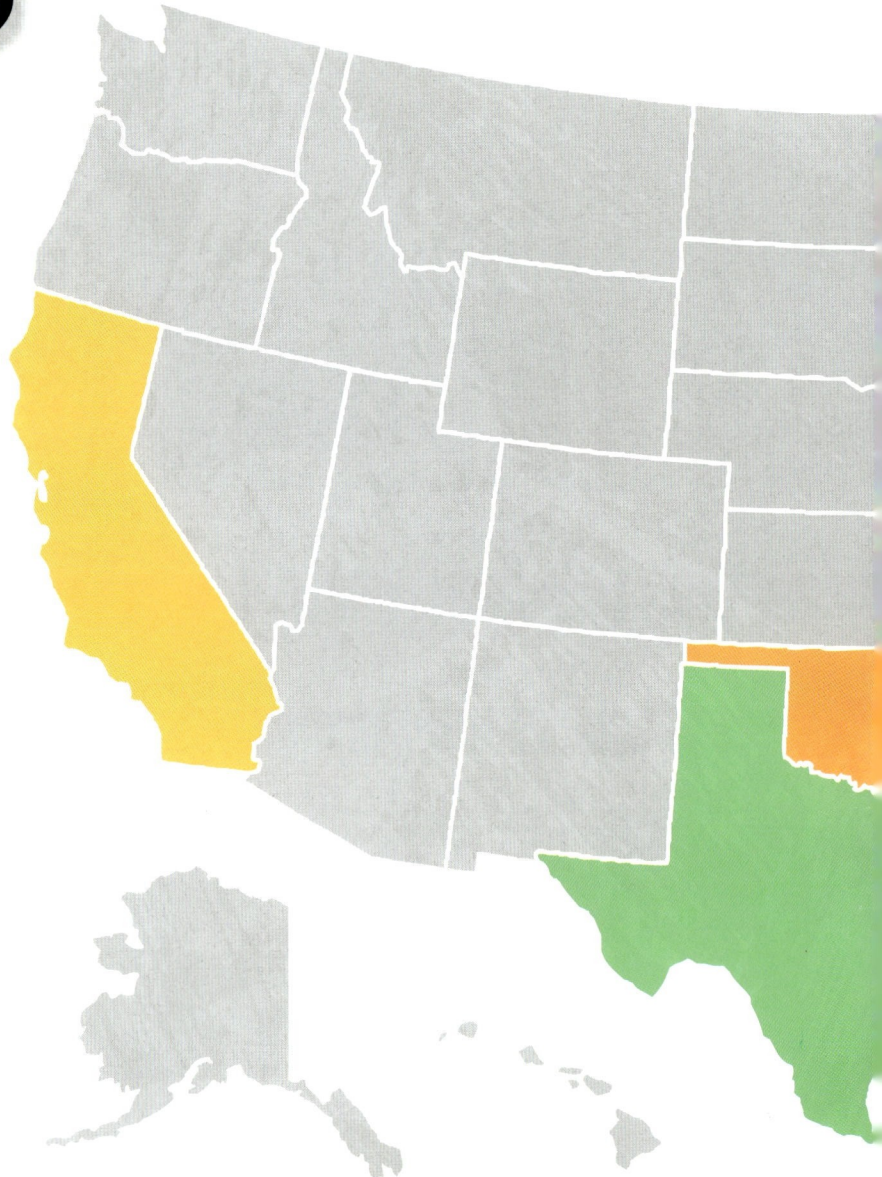

[21ST CENTURY SKILLS LIBRARY]

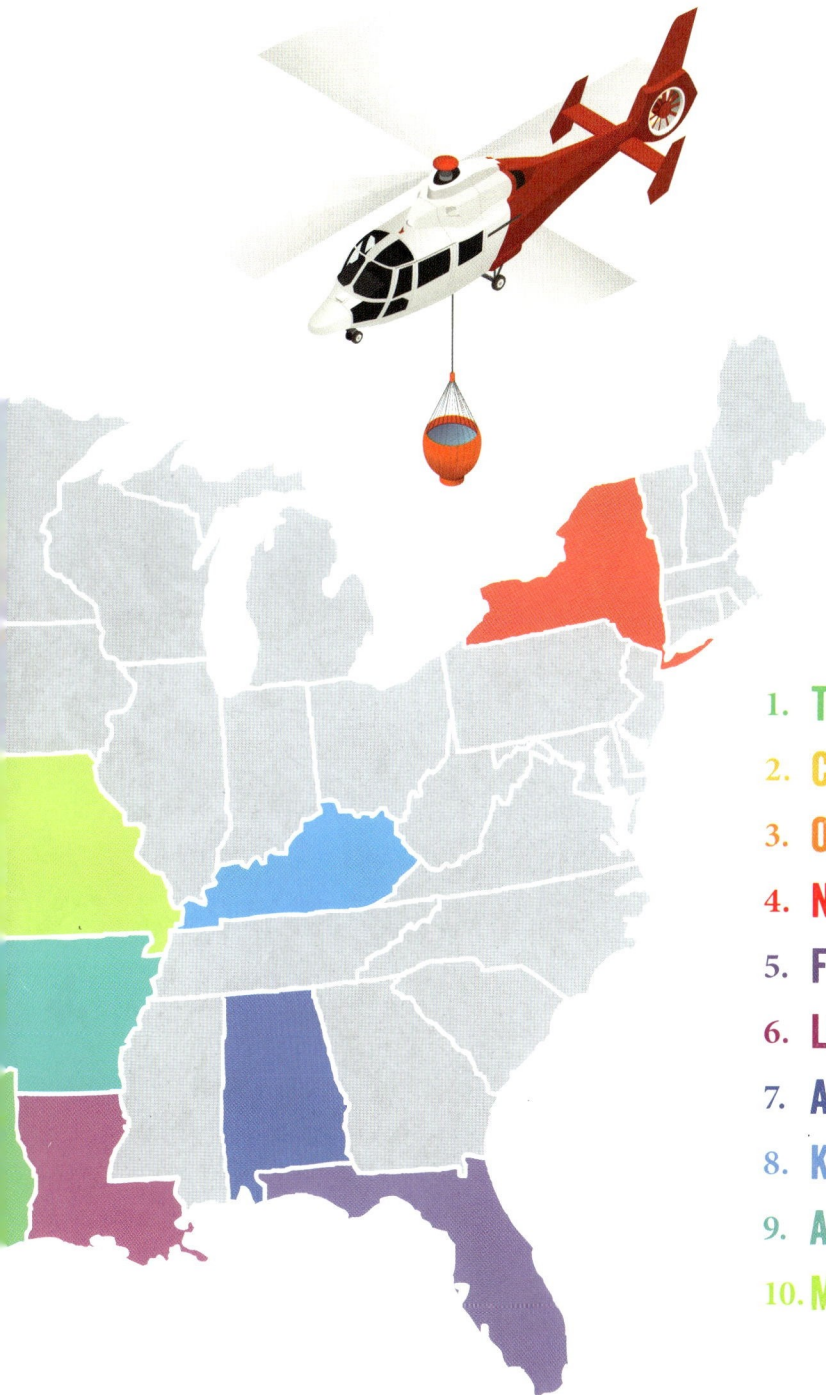

1. **TEXAS**
2. **CALIFORNIA**
3. **OKLAHOMA**
4. **NEW YORK**
5. **FLORIDA**
6. **LOUISIANA**
7. **ALABAMA**
8. **KENTUCKY**
9. **ARKANSAS**
10. **MISSOURI**

2013, NBC News

Certain types of natural disasters occur more often. This chart shows the number of events by type of disaster between 2006 and 2016.

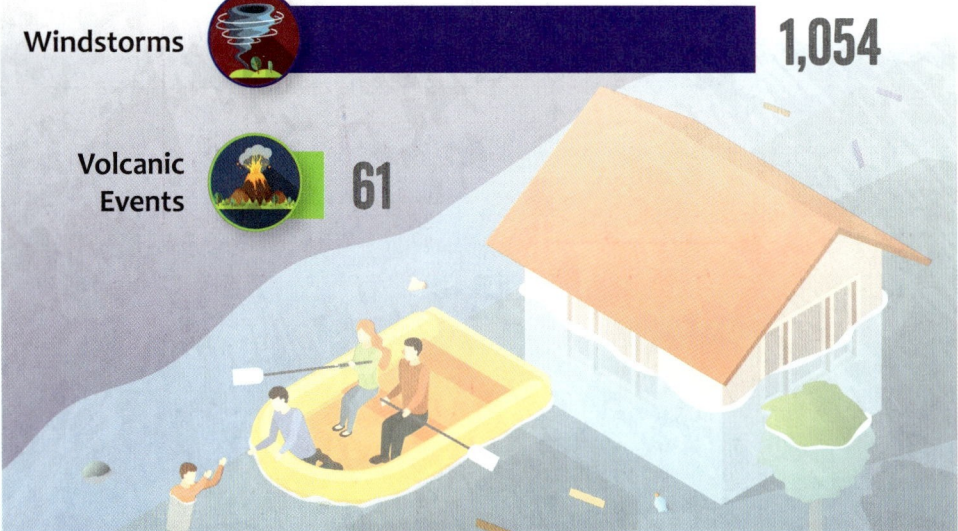

Droughts 273

Earthquakes 290

Floods 1,739

Forest Fires 142

Windstorms 1,054

Volcanic Events 61

2016, Statista

NUMBER OF PEOPLE AFFECTED BY WEATHER-RELATED DISASTERS OVER 20 YEARS (from 1995 to 2015)

Floods

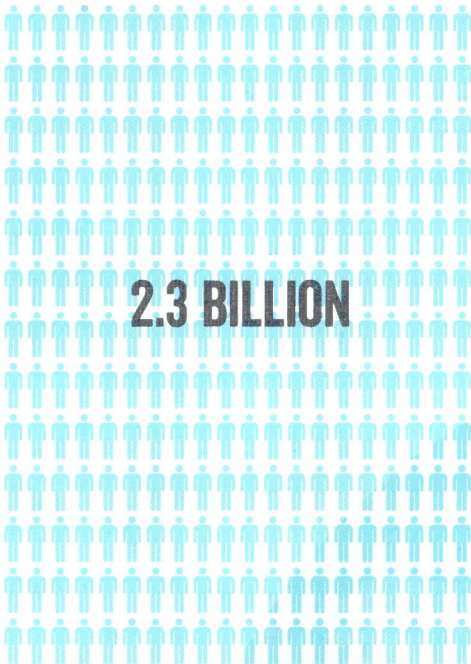

2.3 BILLION

Droughts

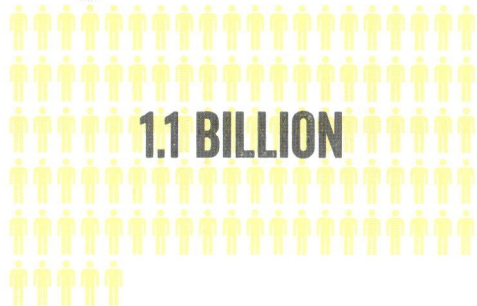

1.1 BILLION

Extreme Temperatures

94 MILLION

Landslides/Wildfires

8 MILLION

Storms

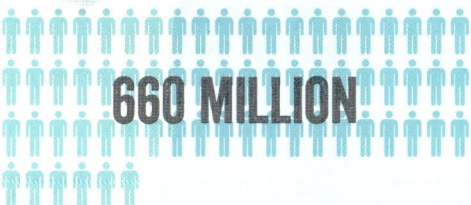

660 MILLION

2015, United Nations Office for Disaster Risk Reduction

Key

 = 10 Million People

How to Plan

Some **continents** have more natural disasters than others. Asia has the most events, while Oceania has the least. These are the numbers of disasters from 2000 to 2009.

Americas
1,334

[21ST CENTURY SKILLS LIBRARY]

Europe
996

Asia
2,903

Africa
1,782

Oceania
169

2012, ERIA

HURRICANES

It is very important to have a plan for natural disasters. About 60 percent of Americans have not practiced what to do in a disaster.

BEFORE

- Have a disaster plan.
- Plan for your pets.
- Have a disaster supplies kit.
- Keep your car filled with gas.
- Keep canned food and bottled water.

DURING

- Stay away from flood areas.
- Stay indoors.
- Leave any home that is not sturdy.
- Go to a shelter if needed.
- Be prepared to **evacuate.**

AFTER

- Stay indoors until told it is safe.
- Be aware of flooding.
- Do not drive through water.
- Stay away from water on foot.
- Don't drink water from the sink.

WATER

14

DECADES WITH THE MOST HURRICANES IN UNITED STATES HISTORY

1940s	1880s	1890s
24	22	21
1910s	**1870s**	**2010s**
21	20	19
2000s	**1930s**	**1850s**
19	19	19
1900s	**1950s**	**1980s**
18	17	15

2020, National Hurricane Center

EARTHQUAKES

Planning ahead for earthquakes is very important because they happen without warning. Damage is **unpredictable** and can be widespread.

BEFORE

- Have a disaster plan.
- Know safe spots in each room.
- Keep a first aid kit.

DURING

- Stay indoors.
- Stay away from windows.
- Do not stand in a doorframe.
- If outside, find a spot away from buildings and trees.
- If you are in a car, stay put.

AFTER

- Make sure there are no **fire hazards**.
- Expect **aftershocks** to be on the way.

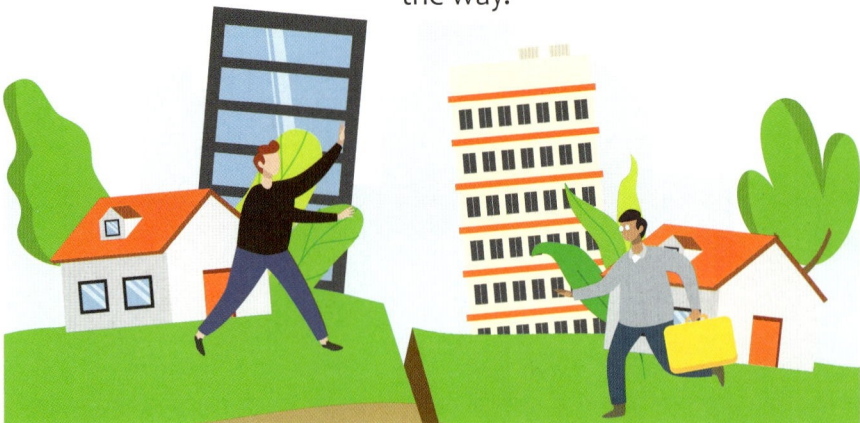

WORST EARTHQUAKES IN U.S. HISTORY

January 26, 1700

Washington, Oregon, and California

1700

Key

= magnitude

9.0

9.0

9.0

1900

April 1, 1946

Alaska

November 10, 1938

Alaska

1950

March 9, 1957

Alaska

March 27, 1964

Alaska

8.2

9.2

2000

8.6

8.6

2014, Amity Insurance

WILDFIRES

In some parts of America, wildfires are a big threat. Some years, as many as 10 million acres (4 million hectares) are burned by wildfires.

BEFORE
Know if you're in a wildfire area.

If you are, have a disaster plan. Plant **fire-resistant** plants around your house. Have a long garden hose.

DURING
Evacuate if told to do so.

Wear protective clothes. Pick an evacuation **route** that is away from danger zones.

AFTER
Stay away from wildfire areas until everything is safe.

Even after the fire appears to be out, danger could still be present.

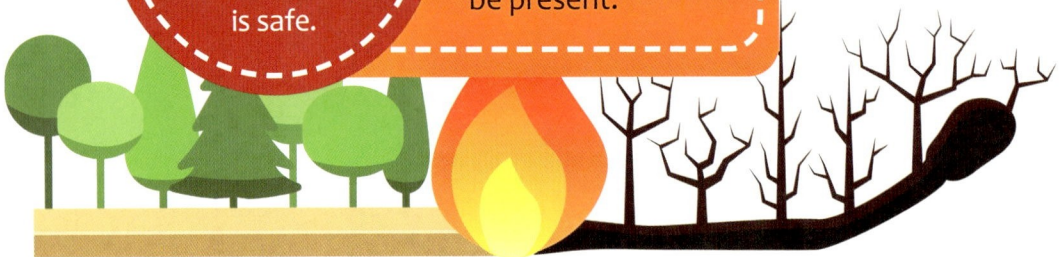

[21ST CENTURY SKILLS LIBRARY]

NUMBER OF WILDFIRES BY YEAR IN THE UNITED STATES

1995	1996	1997	1998
82,234	96,363	66,196	81,043
1999	2000	2001	2002
92,487	92,250	84,079	73,457
2003	2004	2005	2006
63,629	65,461	66,753	96,385
2007	2008	2009	2010
85,705	78,979	78,792	71,971
2011	2012	2013	2014
74,126	67,774	47,579	63,312
2015	2016	2017	2018
68,151	67,743	71,499	58,083

2019, Insurance Information Institute

BLIZZARDS

Nearly every area can be **impacted** by some type of natural disaster. About 80 percent of Americans live in counties that have been hit by a weather disaster. Having a plan for blizzards is very important if you live in a colder state.

❄ Stay indoors.

❄ Eat food. Food provides the body with energy.

❄ Always have a cell phone with you.

❄ Make sure to drive a car with a full gas tank.

❄ Stay dry. Wet clothing can be dangerous.

❄ Tell others where you're going.

MOST EXPENSIVE BLIZZARDS IN U.S. HISTORY

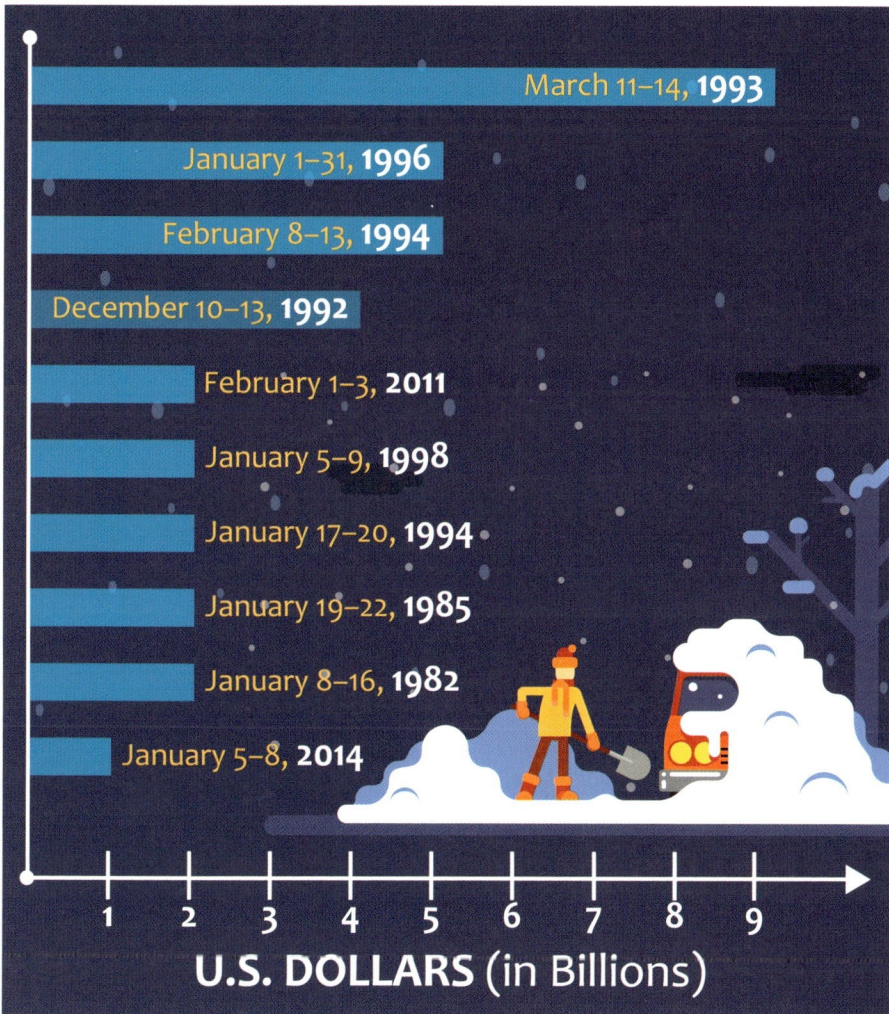

March 11–14, **1993**

January 1–31, **1996**

February 8–13, **1994**

December 10–13, **1992**

February 1–3, **2011**

January 5–9, **1998**

January 17–20, **1994**

January 19–22, **1985**

January 8–16, **1982**

January 5–8, **2014**

1 2 3 4 5 6 7 8 9

U.S. DOLLARS (in Billions)

2015, Wired

TORNADOES

It is very important to know what disasters happen where. Most tornadoes occur throughout the central United States.

1 Get to a basement.

2 If there is no basement, stay under a sturdy piece of furniture.

3 Avoid being in a car or mobile home.

4 If you're outside, find a ditch to lie in.

5 Go to an interior hallway.

[21ST CENTURY SKILLS LIBRARY]

MOST TORNADOES BY STATE
(average per year from 1991 through 2010)

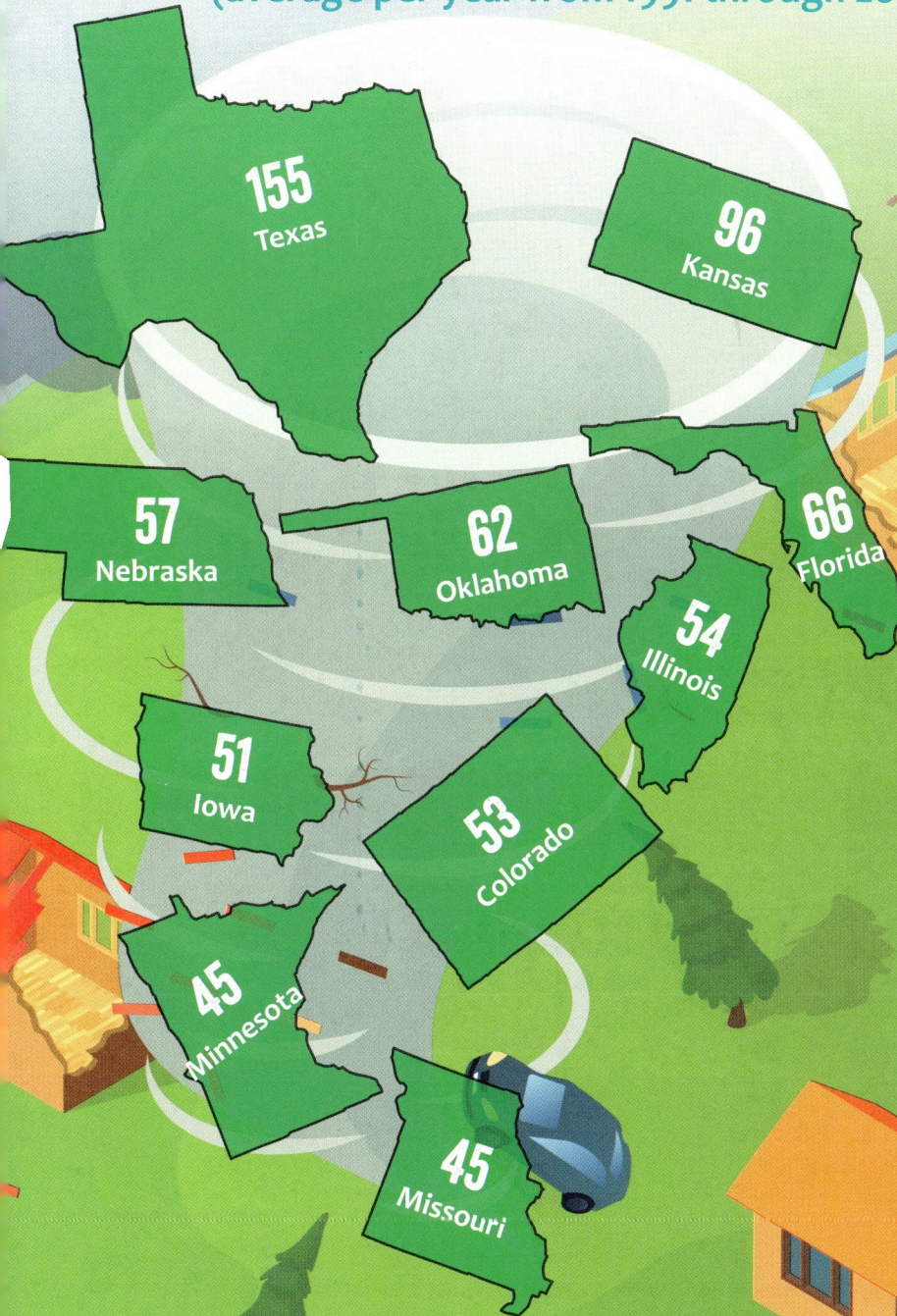

155
Texas

96
Kansas

57
Nebraska

62
Oklahoma

66
Florida

54
Illinois

51
Iowa

53
Colorado

45
Minnesota

45
Missouri

2019, Axios

VOLCANOES

1912
Novarupta
*Alaska Peninsula,
United States*

1980
Mount St. Helens
*Washington,
United States*

1783
Laki
Iceland

2018
Kīlauea
*Hawaii,
United States*

2019
Popocatépetl
*Mexico-Puebla-
Morelos, Mexico*

Volcanoes do not often have large **eruptions**.
But when they happen, the damage can be severe.
These are the largest eruptions in the last 300 years.

2014–15
Bárðarbunga
Iceland

1883
Krakatau
Sunda Strait,
Indonesia

1991
Mount Pinatubo
Luzon,
Philippines

1815
Mount Tambora
Indonesia

2016, Live Science

THE WORST YEAR

An especially bad year for natural disasters was 2017. It was the worst year in recorded history for the United States. Total damages were over $300 billion. The United States had three hurricanes above category 4 hit land. This timeline shows the disasters that caused more than $2 billion in damages that year.

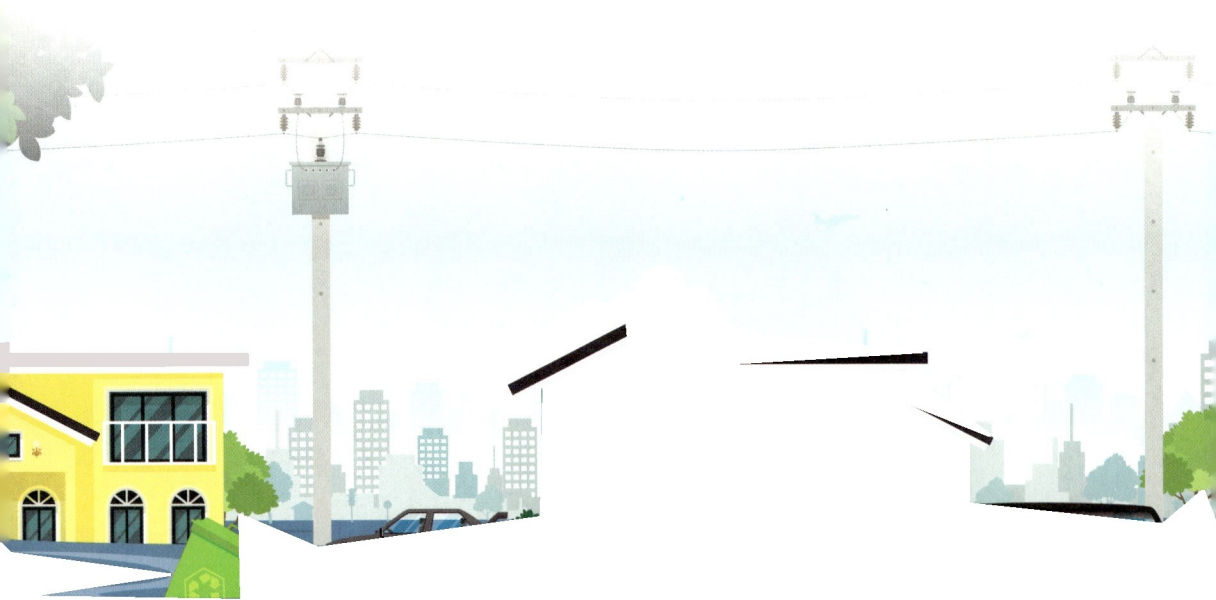

NATURAL DISASTERS IN 2017

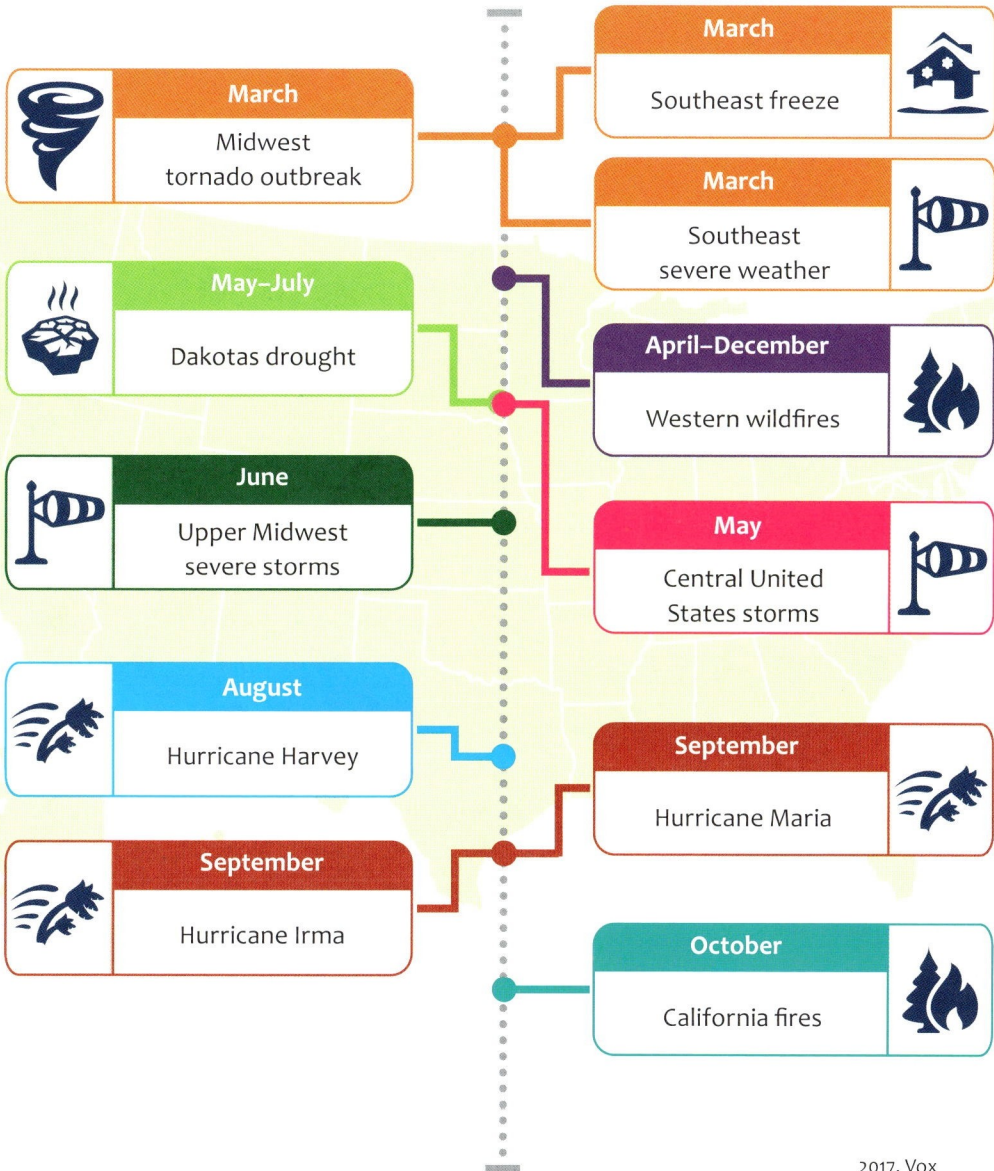

March
Midwest tornado outbreak

March
Southeast freeze

March
Southeast severe weather

May–July
Dakotas drought

April–December
Western wildfires

June
Upper Midwest severe storms

May
Central United States storms

August
Hurricane Harvey

September
Hurricane Maria

September
Hurricane Irma

October
California fires

2017, Vox

Helping Others

GOVERNMENT SPENDING ON DISASTERS BY DEPARTMENT

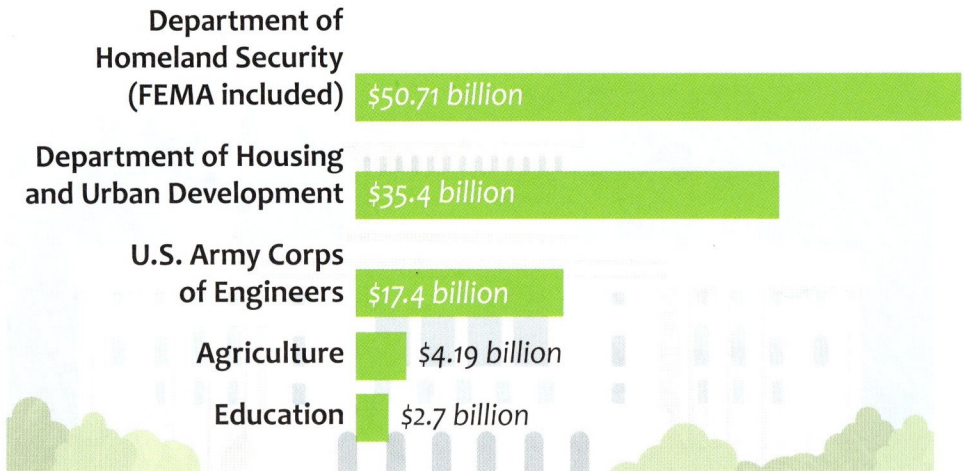

Department of Homeland Security (FEMA included) — $50.71 billion

Department of Housing and Urban Development — $35.4 billion

U.S. Army Corps of Engineers — $17.4 billion

Agriculture — $4.19 billion

Education — $2.7 billion

2018, University of Pennsylvania

Even when a natural disaster does not impact your area, you can still help. Many people will lose their homes or belongings in a bad disaster. There are steps that can be taken to help.

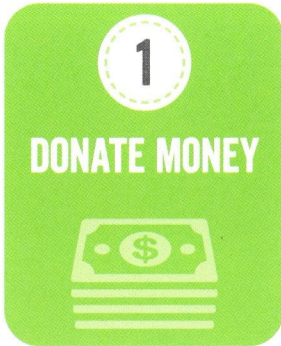

1

DONATE MONEY

- People need money to recover.
- There are many organizations that help get money to people in need.
- Money can be raised in fundraisers.

- Volunteers can help clean up.
- Volunteers can help rebuild structures after a disaster.
- Volunteers can help feed people in need.

2

VOLUNTEER

3

DONATE SUPPLIES

- Food and other supplies are needed after a disaster.
- Clothes can also be useful for people who were impacted by a disaster.
- Check with volunteer organizations to see what supplies are needed.

Activity

BUILD A DISASTER KIT

A great way to be prepared for natural disasters is to have a disaster kit. Disaster kits are simple to put together.

Collect the following items:

- Bottled water
- Dried foods
- Radio
- Canned goods
- Cash
- Toiletries
- Can opener
- First aid kit
- Whistle
- Contact list
- Flashlights

Store the kit in a place where it is safe and easy to get to.

Learn More

BOOKS

Goin, Miriam. *National Geographic Readers: Storms!* Washington, DC: National Geographic Kids, 2009.

Griffey, Harriet. *Earthquakes and Other Natural Disasters.* New York, NY: DK Children, 2010.

Hayes, Amy. *Meteorology and Forecasting the Weather.* New York, NY: PowerKids Press, 2018.

Van Rose, Susanna. *Volcano and Earthquakes.* New York, NY: DK Eyewitness Books, 2014.

WEBSITES

National Geographic Kids
https://www.natgeokids.com/za/tag/natural-hazards-and-disasters

Generation Genius
https://www.generationgenius.com/natural-disasters-for-kids

Easy Science for Kids
https://easyscienceforkids.com/earth-science/natural-disasters

BIBLIOGRAPHY

Live Science. "10 Deadliest Natural Disasters." April 4, 2018. https://www.livescience.com/33316-top-10-deadliest-natural-disasters.html

National Hurricane Center. "U.S. Hurricane Strikes by Decade." August 1, 2005. https://www.nhc.noaa.gov/pastdec.shtml

Wharton University of Pennsylvania. "Federal Disaster Rebuilding Spending: A Look at the Numbers." February 22, 2018. https://riskcenter.wharton.upenn.edu/lab-notes/federal-disaster-rebuilding-spending-look-numbers

NBC News. "10 States with the Most Natural Disasters." May 27, 2013. https://www.nbcnews.com/businessmain/10-states-most-natural-disasters-6C10088195

GLOSSARY

aftershocks (AF-tur-shoks) small earthquakes that come after a large one

continents (KAHN-tuh-nuhnts) the seven large land masses on Earth, such as North America and Asia

eruptions (i-RUHP-shuhn) the sudden explosions of rock, hot ash, and lava from a volcano

evacuate (i-VA-kyoo-ayt) to leave a dangerous area and go to a safe place

fire hazards (FIRE HAZ-urds) items or systems that could start a fire, such as downed power lines

fire-resistant (FIRE ri-ZISS-tuhnt) very difficult to burn

impacted (im-PAK-tid) changed by a major event or influence

route (ROOT) the particular course followed to get somewhere

severe (suh-VEER) very strong; able to cause a lot of damage

unpredictable (pri-DIKT-uh-buhl) impossible to be known beforehand

vulnerable (VUHL-nur-uh-buhl) open to being attacked, hurt, or damaged

INDEX

ABOUT THE AUTHOR

Alexander Lowe is a writer who splits his time between Los Angeles and Chicago. He has written children's books about sports, technology, science, and media. He has also done extensive work as a sportswriter and film critic. He loves reading books of any and all kinds.

32

[21ST CENTURY SKILLS LIBRARY]